—OFFICE OF—

MINONG COPPER COMPANY,

22 WALKER BLOCK,

Detroit, June 23, 1883.

To ....

Dear Sir:

You are hereby notified that an assessment of fifty cents on each share of the stock of this Company was made and levied by the Directors this day, payable at the office of the Company on the twenty-third day of July next.

Yours, etc.,

....

Secretary and Treasurer.

OFFICE OF MINONG COPPER CO.,
22 WALKER BLOCK,
Detroit, June 28th, 1883.

To the Stockholders of the
Minong Copper Company:

At a late meeting of the Board of Directors of this Company, it was concluded, upon taking into consideration the present value of copper and the future prospects thereof, to suspend the working of the mine of this Company for the present.

This conclusion has not been arrived at for the reason that the property is not valuable, nor that the indications would not justify a further expenditure of money. The operations of the past have been in the nature of an experiment, with the hope of striking such a body of ore as would justify confining the works exclusively to the recovery thereof. The property is, by experts who have examined it, and by experienced miners, declared to be valuable, and it only wants the money capital to develop it.

In all the works ore has been discovered, indicating that with the prosecution thereof it would lead to a success; but even with it, at the present price of copper the Board of Directors are not disposed to continue expenditures until it shall appear a satisfactory result will be accomplished in the future.

The Directors have therefore levied an assessment of fifty cents per share, for the purpose of relieving the property until it shall be determined by the Stockholders hereafter what shall be their desire as to further operations.

This will leave in the treasury a fund sufficient to pay the taxes on the property and other incidental expenses for several years.

A statement of the affairs of the Company to date is sent herewith.

Very Respectfully,

E. W. HUDSON,
Secretary.

TREASURER'S REPORT TO THE STOCKHOLDERS OF THE MINONG COPPER CO. FROM ITS ORGANIZATION TO JUNE 23, 1883.

RECEIPTS.

Subscription for 20,000 shares stock	$20,000 00
Assessment No. 1, of 50c. per share	10,000 00
" No. 2, of 50c. "	10,000 00
" No. 3, of 75c. "	15,000 00
Sales of Copper	13,908 57
Rents of Houses at the Works	1,011 00
Doctor's Fees	610 75
Team Work and Sales of Wood	501 80
Cash Rec'd on Notes of Company outstanding	3,800 00
	$74,832 12
Balance	5,959 35
	$80,791 47

DISBURSEMENTS.

Paid for Property, including Merchandise	$20,000 00
" Interest, Postage, Books, Printing, Stationery, Lithographing, etc.	666 60
" Insurance	1,248 76
" Secretary's Salary	2,316 67
" Sundry Trips to Works and Ft. and Mdse.,	2,980 12
" Taxes and Office Rent	542 19
" Iron, Steel, Coal and Lumber	1,128 96
" Powder	1,945 17
" Smelting and Transportation of Copper	929 37
" Mining Captain's Wages	3,452 00
" Doctor's Wages (acting as Clerk)	3,517 32
" Miners' Wages, per month,	31,470 93
" " " contract	2,490 37
" Chopping Wood	1,303 01
Bills Payable to Date	3,800 00
Estimated Due the Men and Interest and Expense,	3,000 00
	$80,791 47

E. W. HUDSON,

Treasurer.

Secretary & Treasurer
Jan 23/83

Minong Mining Company,

Organized December 16th, 1874,

UNDER THE

General Mining Laws of the State of Michigan.

CAPITAL $1,000,000.

Shares Twenty-five Dollars Each.

DIRECTORS.

STANLEY G. WIGHT,	Detroit, Mich.
A. W. COPLAND,	Detroit, Mich.
C. M. GARRISON,	Detroit, Mich.
JOHN BELKNAP,	Detroit, Mich.
GEO. W. GILBERT,	Detroit, Mich.
W. M. McCONNELL,	Pontiac, Mich.
WM. H. STONE,	St. Louis, Mo.

President.

STANLEY G. WIGHT.

Vice President.

A. W. COPLAND.

Secretary and Treasurer.

C. M. GARRISON.

Office of the Company:

No. 160 Jefferson Avenue, Detroit.

PROSPECTUS

OF THE

Minong Mining Company.

1875.

Office No. 160 Jefferson Avenue, Detroit.

DETROIT:
DAILY POST BOOK AND JOB PRINTING ESTABLISHMENT.

1875.

MINONG MINING COMPANY.

The property of this company consists of 1,455 acres of mineral lands, in sections 22, 23, 26, 27, 34 and 35, township 66 north, range 35 west, near McCargoe's Cove, on the north side of Isle Royale, Lake Superior, as will be seen by the accompanying map. The lands are traversed by heavy metaliferous belts and transverse veins carrying copper. showing very extensive ancient mine works on both. The property has been very thoroughly explored the past season, and discoveries have been made that we believe to be of greater value and importance than any heretofore made on Lake Superior.

In presenting the advantages of this property to the public the Directors have no hesitancy in recommending it, fully believing it possesses all the requisites to make a rich and valuable copper mine, one that must be a paying investment.

The ancient mines found on this property far exceed in richness and magnitude any heretofore discovered and in order to give some idea of the importance ancient mining work bears as a guide to the discoveries of rich mineral lodes, we call attention to the following extracts from early history of the island and scientific reports in relation to the subject.

The discovery of native copper naturally excited the wonder of the first voyageurs, and their references to it are numerous. In the explorations of Lake Superior, in 1669-70, the Jesuit missionary, Father Dablon says: "Towards the west, on the same north shore, is the island most famous for copper called Minong ("the good place") Isle Royale. This island is twenty-five leagues in length; it is seven leagues from the main land, and sixty from the head of the lake. Nearly all around the island, on the water's edge, pieces of copper are found mixed with pebbles, but especially on the side which is opposite the south, and principally in a certain bay which is near the northeast exposure to the great lake."

In speaking of the ancient mines, Prof. J. W. Foster, in his late work on the Pre-Historic Races of America, says: "The high antiquity of this mining is inferred from these facts: That the trenches and pits were filled even with the surrounding surface, so that their existence was not suspected until many years after the region had been thrown open to active exploration; that upon the piles of rubbish were found growing trees which differed in no degree, as to size and character, from those in the adjacent forest, and that the nature of the materials with which the pits were filled, such as a fine washed clay enveloping half decayed leaves, and the bones of such quadrupeds as the bear, deer and caribou, indicated the slow accumulation of years, rather than a deposit resulting from a torrent of water."

Prof. Whittlesey, in his report to the Smithsonian Institute on Ancient Mining, on Lake Superior, in

1862, says: "The first public announcement, so far as we are aware of the remains of ancient mines in the copper region, is that by Mr. S. O. Knapp, agent of the Minnesota Mining Company, in 1848. Dr. Charles T. Jackson brought forward the subject in his Geological Report to the United States Government in 1849, and gave some interesting details of what had been discovered up to that time. Further mention of it was made by Messrs. Foster and Whitney, in their report in 1850, and several illustrations were given. Since then our knowledge of the subject has been very much enlarged by the prosecution of mining operations on the very sites of the ancient works. It must not, however, be supposed that our information is complete. It is by no means an easy task to discover remains, buried as those of the ancient mines of Lake Superior are, in extensive and dense forests, where the explorer can only see a few rods, or, perhaps, yards, around him, and where there is seldom anything which rises sufficiently high above the surface to attract the eye.

They are, for the most part, merely irregular depressions in the soil, trenches, pits and cavities, sometimes not exceeding one foot in depth, and a few feet in diameter. Thousands of persons had seen the depressions prior to 1848, who never suspected that they had any connection with the arts of man, the hollows made by large trees overturned by the wind being frequently as well marked as the ancient excavations. Besides this there are natural depressions in the rocks on the outcrop of veins formed by the decomposition of the minerals, that resemble the troughs of the ancient miners, as they appear after the lapse of centuries.

"There is not always a mound or ridge along the side of the pits, for most of the broken rock was thrown behind, nearly filling up the trenches.

"A mound of earth is as nearly imperishable as any structure we can form. Some of the tumuli of the west retain their form and even the perfection of their edges at this day. But mere pits in the earth are rapidly filled up by natural processes. Some of those which have been re-opened, and found to have been originally ten feet deep, are now scarcely visible; others that have a rim of earth around the borders, or a slight mound at the side, and were at first very shall low, are more conspicuous at present than deep ones without a border.

"There are, however, pits of such size as could not fail to surprise one at first view, were not the effect destroyed by the close timber and underwood with which they are surrounded.

"A basin-shaped cavity, 15 feet deep and 120 feet in diameter, would immediately attract the eye of the explorer were it properly exposed. But it is not unusual to find ten and twelve feet of decayed leaves and sticks filling a trench, and no broken rock or gravel.

"In such cases, a fine red clay has formed towards the bottom a deposit from water, which indicates the long period of time since the excavation was made. Although the old works are not always situated upon what would be considered good veins, yet they are regarded by practical miners as pretty sure guides to valuable lodes.

"In the opening of our principal mines we have followed in the path of our ancient predecessors, but with much better means of penetrating the earth to great depths. The old miners performed the part of surface explorers."

Mr. Henry Gillman, of the United States Lighthouse Department, in an article on the Ancient Mine work at Isle Royale, Lake Superior, published in *Appleton's Journal*, Aug. 9th, 1873, says: "The remarkable discoveries which have recently been made at Isle Royale, Michigan, of the mining and other works of an ancient people are entitled to a more general recognition than for various reasons they have as yet received. Owing to a visit lately made the island, and through the kindness of a gentleman well known in mining interests, who is at present engaged in developing the mineral resources of the place, the writer has become acquainted with something of the wonderful extent and character of the works referred to.

"Though it is probable that not one-tenth of those ancient excavations have so far been revealed, some idea of their extent may be arrived at, from the statement of the gentleman mentioned, who has calculated that, at one point alone on three sections of land toward the north side of the island, the amount of labor performed by those ancient men far exceeds that of one of our oldest copper mines on the south shore of Lake Superior, a mine which has now been constantly worked with a large force for over twenty years.

"At a deep inlet, known as McCargoe's Cove, on the north side of the island, excavations such as are described extend in almost a continuous line for more than two miles, in most instances the pits being so close together as barely to permit their convenient working. The stone hammers, weighing from ten to even thirty pounds, the chief tool with which the labor was performed, have been found in cart loads. They are either perfect, or are broken from use, and the fragments of large numbers of them are found intermingled with the *debris* on the edge of the pits, or at their bottom.

"The pits which have been examined by being cleaned out invariably had at top a large deposit, mostly of vegetable remains, the accumulations of many a fall-of-the-leaf, beneath which lay a thick bed of charcoal and mud, mingled with fragments of copper-bearing rock. The method of mining pursued by those people was evidently, on turning back the overlying drift, to heat the rock through the aid of fire; then, when by the application of water the rock was sufficiently disintegrated, to attack and separate it with their great hammers. The works, generally pits of from ten to twenty feet in diameter, and from twenty to sixty feet in depth, wherever examined, being sunk through the few feet of superincumbent drift, where it exists into the amygdaloid copper-bearing rock. They invariably are on the richest veins, and the intelligence displayed in the tracing and following of the veins when interrupted, etc., has elicited the astonisnment of all who have witnessed it, no mistakes having apparently been made in this respect.

"An experienced mining captain computed that two hundred of these men with their rude methods, could barely be equivalent to two of our skilled miners. Though no exact estimate can now be made as to the length of time occupied in the prosecution of those extensive works, it does not seem too much to estimate hundreds of years for their accomplishment."

We believe no spot in the Lake Superior country yet discovered shows anything like the amount seen on this property. This was undoubtedly the great mining centre of the ancients. The metaliferous belt and veins are elevated above the level of the surrounding country, so that good drainage can easily be had. The descending grade from any part of the property to the cove is easy, and there is good soil for building roads. No road, however, will be required to exceed one and one-half miles in length. The property has an abundance of good timber, well adapted for building purposes and the general uses of a mine. There is no great amount of expensive work to be performed to develop this mine, as nature has done so much for it.

The harbor (McCargoe's Cove) is a bay running into the land two and one-half miles, with an abundance of water for the largest class of vessels. Its shores are rocky and bold, and a very moderate expense would prepare docks for a very large business.

The location presents many features that are extremely inviting to all who are acquainted with mining. The accessibility of the harbor to the points to be mined, the timber, the drainage, and, best of all, the evidence on all sides of rich and productive copper-bearing ground, make it an offer seldom presented to the public.

With these facts in view, it is confidently believed that there have been few discoveries in this or any other country which have presented such an assured promise of a rich and valuable copper mine as this, and the working of it will be vigorously prosecuted by the Directors on the opening of navigation.

The accompanying letters, from persons who have visited the location and examined the property, are entitled to consideration. They are from men distinguished for their scientific and practical acquaintance with mines and mining, and for the candor and caution with which they are accustomed to speak on questions of this character:

FRANKLIN MINE, Oct. 30, 1874.

P. A. Hitchcock, Esq., Pontiac, Mich.:

Dear Friend—Your favor of the 26th came all right, and, in reply, I would say that I visited the McCargoe's Cove property on Isle Royale, you speak of, last July, and all I can say about it is that there is more extensive Indian diggings on the property than I have ever seen on the south shore, put them all together. Now you know the Indians never dug without they got copper; therefore, I think there must be a large amount of copper on the property. In fact, it is a better surface show than I have ever seen anywhere. I am not interested in the property one cent, directly or indirectly. My opinion is entirely disinterested. Yours truly,

R. UREN.

ISLAND MINE, ISLE ROYALE, LAKE SUPERIOR, Mich.,
Dec. 4, 1874.

Stanley G. Wight, President Minong Mining Co.:

Dear Sir—Late in September last I embraced an opportunity to visit McCargoe's Cove, Isle Royale, and it is with feelings of great pleasure I now give you my impressions of the value of your property. I was more than pleased to find all previous reports fully sustained, and may say with truth the half had not been told. The first thing particularly attracting the attention is the vast amount of ancient mine work done. It is not only surprising, but absolutely wonderful. It is well known that the ancients wasted little time in barren ground, and, judging from amount of work done, this favored spot must have been, centuries ago, the headquarters of the ancient miners, and, in all probability, was the Calumet and Hecla of that distant period. By their primitive method of mining, and with their rude implements, thousands of men would seem to have been employed for years, and the amount of copper obtained must have been very large. I notice the belts running with the formation south 60° west magnetic are very wide, and the fine mass of copper taken out this fall strengthens my opinion that in these belts will be found large deposits of mass, barrel and stamp copper.

I observed with great interest the number of veins cutting the formation (course about south 30° west magnetic), and the extent, in length and depth, of the ancient excavations upon them is wonderful. I very much regret that my visit was confined to one day. Your property presents so many promising indications of value, and possesses so many peculiar features of interest, that I could profitably spend a whole week in examining and studying it up.

The location is admirably situated for mining operations. It is well timbered, and of a quality useful for mining purposes. You have also a safe and beautiful harbor.

In conclusion, allow me to congratulate you on the acquisition of such a valuable mining property.

Very respectfully yours,

GEORGE HARDIE.

MARSHALL, Dec. 19, 1874.

Stanley G. Wight, Esq., President Minong Mining Co.:

Dear Sir—I am well acquainted with the Minong Mining Company's property, embracing 1455 acres, on the west side of McCargoe's Cove, Isle Royale. The property is well covered with timber suited to the requirements in and about mines; the surface is accessible in all parts. McCargoe's Cove, a good and safe harbor, is contiguous. The rock formation is the well-known copper bearing system of Keweenaw and Houghton counties, of the south shore of the lake; this rises in ridges from fifty to three hundred feet, making drainage good and easy in nearly all parts of the property.

Commencing at the harbor in section 26, there is a ridge of bedded diorite having a direction south 60° west; its length in this property is something more than one mile. This ridge is traversed by veins, the greatest number of which bear south 10° to 30° west; there are more than twenty in this property, the greatest number of them have been extensively mined by a pre-historic race.

The depth of some of these old mines has been found to be sixty feet; where the veins have been uncovered, the excavations in them by the old miners, when cleaned out they have been found well filled with copper; one thousand feet west of the harbor, in Sec. 26, west half, there is a group of several veins; the works of the old miners mark their course more than twelve hnndred feet, the veins are united in the north part of the ridge. The evidences are that these mines will pay to mine. They are of good width; other and large veins near this group that have been extensively mined have been found.

The Witthaus veins in the east half Sec. 27, form another group in which the works of the old miners are as extensive as even the works of this kind in both Minnesota and National Mines, on the south side of the lake; these veins are large and well defined, and will, I believe, be found to be of great value when mined.

Flanking this ridge before mentioned, on the south side, there is a belt or bed of highly metamorphosed sedimentary deposits, the thickness of which where it has been seen, and when it is not covered with drift, varies from twenty to forty feet. This bed contains the largest amount of ancient mine work that has ever been found on bed, vein or locality. These old works in this bed, in Section 27, have a continuous length of one-half of a mile; it was in one of these old excavations, in this locality, and in this bed the large mass of copper was found that is now in the city of Detroit, that has attracted so much attention. In Section 26, in this property, where this bed again comes to the surface, there are other and extensive works of these old miners wherever it has been uncovered, and when the old works have been cleaned up, it has been found well filled with copper. This deposit of copper appears to be more extensive than any other ever discovered, not excepting the famous Calumet and Hecla, on the south shore of the lake. Other beds containing copper contiguous to and parallel with this have been found there to contain works of the old miners.

These grounds are well located to be easily and economically mined; one road, either wagon or rail, over an easy and even grade, will command all the veins and the bed deposits that will be required to be mined. The surface advantages are not excelled or equaled by those of any other mining property that I have seen.

It has fallen to my lot to have seen all of the mines in their earliest stages of development, and in none of them have I had as much confidence as I have in this. The results of mining this property connot be doubtful, and must be largely remunerative to the shareholders.

Respectfully yours,

S. W. HILL.

OSCEOLA CONSOLIDATED MINING CO.,
HOUGHTON COUNTY, MICH.,
CALUMET, Dec. 22, 1874.

Mr. Stanley G. Wight, President Minong Mining Co.:

My Dear Sir—In company with several others, I visited the property lying adjacant to McCargoe's Cove, on Isle Royale, during the month of October, 1874.

We found a land-locked harbor, and in examining the property we found the most extensive ancient diggings I have ever seen, the somewhat noted pits and diggings upon the old Minnesota Mine, and in Ontonagon County, being a mere trifle as compared with what we saw upon your property. We had not the time or appliances to unwater the many pits, but saw where work had been done extensively upon some seven veins. Following the course of pits westwardly to section 27, we came upon a very extensive belt, and lying in a pit, and upon it was the mass of copper taken to Detroit by Mr. A. C. Davis, and the evidence was conclusive that this mass was a part of and mined from this belt. The ancient mining upon this belt was also very extensive, which, taken with the fact long ago developed in Ontonagon County, that in most cases this mining was done in the vicinity, and usually in contact with copper in quantity, gives of itself a warrant of promise to investors in the enterprise.

And, in conclusion, I can say the evidences already developed upon your property are sufficient to warrant capital in still further following up those developments.

Very truly yours,

F. G. WHITE.

Stanley G. Wight, President Minong Mining Co.:

Dear Sir—I spent some eight weeks the past summer in examining the property of the Minong Mining Company at McCargoe's Cove, Isle Royale, Lake Superior.

The property is wholly on the Trap Range, and the rocks are the same as seen in Keweenaw County, south shore of the lake.

The harbor (McCargoe's Cove) is all that could be desired, having abundant depth of water, and room for any number of vessels. The expense of preparing docks to do your business will be comparatively small, as twenty-seven feet from the shore you have ten to twelve feet of water.

The timber on the property is pine, tamarac, cedar, balsam, poplar and white birch. There is pine enough to make all the lumber that will be needed for a number of years. The balance of the timber is well adapted to the wants of a mine.

The property to be mined is elevated above the Cove, which makes drainage easy; the grades and soil for roads are good.

The property has a number of transverse veins that show a large amount of ancient mine work. I cleaned out one of the ancient pits on a vein near the east line of section 27. Its depth was twenty-six feet. I found a vein two feet wide. I put in a blast, and the result was about forty pounds of barrel copper, beside stamp work. I consider this vein a valuable one to mine for copper, and the chances of getting a paying mine I regard as almost certain.

The property has a sedimentary belt running through it that is highly metaliferous. This belt has also been extensively mined by the ancieut miners. I cleaned out three of the ancient pits on this belt, distant one from the other nearly a mile, in all of which I found barrel and stamp copper in quantity to warrant mining. At one point I made a cross cut across this belt, and found it contained copper for a width of forty feet. It was in this cross cut I found the mass estimated to weigh 6,000 pounds. This mass had been detached from its bed by the ancient miner. I found a number of pieces of copper besides the mass, weighing from an ounce to seventeen pounds. I did not do any blasting in the cross cut, but broke the rock with a pick in many places, and invariably found it well fllled with copper.

In conclusion, I regard the property as extremely valuable for mining, and have no hesitancy in giving it as my candid opinion that, when mined, it will make a dividend-paying copper mine.

Respectfully yours, A. C. DAVIS.

www.ingramcontent.com/pod-product-compliance
Lightning Source LLC
LaVergne TN
LVHW020637110826
845149LV00004B/1249
* 9 7 8 1 4 1 8 1 9 1 0 8 5 *